Mehdi Boukil

Experiential Tourism in the Historic Medina of Fez

Mehdi Boukil

Experiential Tourism in the Historic Medina of Fez

From Immersive Guided Discovery to Treasures of Craftsmanship and Gastronomy

ScienciaScripts

Imprint
Any brand names and product names mentioned in this book are subject to trademark, brand or patent protection and are trademarks or registered trademarks of their respective holders. The use of brand names, product names, common names, trade names, product descriptions etc. even without a particular marking in this work is in no way to be construed to mean that such names may be regarded as unrestricted in respect of trademark and brand protection legislation and could thus be used by anyone.

Cover image: www.ingimage.com

This book is a translation from the original published under ISBN 978-620-6-71600-6.

Publisher:
Sciencia Scripts
is a trademark of
Dodo Books Indian Ocean Ltd. and OmniScriptum S.R.L publishing group

120 High Road, East Finchley, London, N2 9ED, United Kingdom
Str. Armeneasca 28/1, office 1, Chisinau MD-2012, Republic of Moldova, Europe
Printed at: see last page
ISBN: 978-620-7-88298-4

PREFACE BY THE DIRECTOR OF CITE DE METIERS ET COMPETENCES (CMC)

The medina of Fez, with its rich cultural and historical heritage, offers visitors immersive tourism experiences, from discovering traditional crafts to taking part in Moroccan cookery workshops. This experiential tourism allows travellers to immerse themselves in local life, learn the ancestral techniques of craftsmen and taste dishes prepared according to recipes handed down from generation to generation.

This approach finds a particular echo in the skills-based training method advocated by the Office de Formation Professionnelle et de Promotion du Travail (OFPPT). Indeed, the practical training in gastronomy and crafts, centred on the acquisition of concrete, applicable skills, bears striking similarities to the tourist experiences on offer in the medina of Fès. Apprentice chefs and craftsmen benefit from immersive training that plunges them into the heart of their trade, just like the tourists who discover and actively participate in local cultural practices.

The book "Experiential Tourism in the Medina of Fez", written by an experienced tourism management trainer, Travel Agency option, explores this intersection between tourism and practical training using the competency-based approach. It demonstrates how tourists' experiences can be enriched by an in-depth understanding of craft and gastronomic techniques, while highlighting the importance of practical training in preparing professionals to excel in their field.

INTRODUCTION TO BOOK

Experiential tourism is characterised by its focus on immersive, interactive and authentic experiences for travellers. Rather than simply observing tourist sites, visitors actively participate in activities that allow them to discover and understand local culture, traditions and lifestyles. This can include interactions with locals, craft workshops, cooking demonstrations, personalised guided tours, and other experiences that enrich their understanding of the place they are visiting.

The growing importance of experiential tourism in the modern tourism industry stems from a number of key factors:

- Search for authenticity: Travellers are looking for more authentic and meaningful experiences that go beyond standard tourist attractions.

- Personal involvement: Experiential tourism allows visitors to become actively involved in their travel experience, creating more lasting and satisfying memories.

- Local economic impact: It promotes local economic development by supporting local craftsmen, guides and other service providers.

- Competitive differentiation: Destinations that offer unique experiential experiences can stand out in the tourism market and attract travellers looking for memorable and enriching experiences.

The city of Fez, one of the oldest Islamic and Arab capitals in North Africa, is a living testimony to the last twelve centuries of Moroccan history. It has captured the attention of the kings and dynasties who have shaped Morocco's history. the kingdom's destiny, occupying a central position in its historical narrative. Over the centuries, its diverse population has helped to shape a harmonious and unique heritage.Fez was chosen as a case study because of its Medina, listed in its entirety as a World Heritage Site by UNESCO, which is home to a plethora of historic monuments. Among them, Al Karaouiyine, founded in 857, is the oldest university in the Arab world. The Medina of Fez is a symbol of exceptional historical and heritage wealth, offering a diverse exploration of its many facets, including :

• A variety of historic monuments (mosques, medersas, mausoleums, fondouks, palaces, riads, etc.)

• Authentic gastronomy and a distinctive culinary art

• A wide range of crafts (pottery, tanning, brasswork, woodwork, herbalism, etc.)

• Preserved traditions, customs and daily life...

As you explore the medina of Fez, every architectural detail becomes an invitation to immerse yourself in the vibrant history of this imperial city. These elements are not just relics of the past, but living witnesses to a culture that continues to evolve while preserving its deep roots.Historic buildings stand proud on every corner, their facades decorated with intricate designs and architectural details that reflect Islamic art and the influence of the dynasties that have shaped the city. Carved doors and finely crafted wooden balconies are testimony to the meticulous craftsmanship and aesthetic taste of the Fassis people down the ages.This book explores in depth three key aspects that make Fez an essential destination for experiential tourism:

• **Chapter 1: Crafts - The Foundation of the Experience** Discover how traditional crafts such as pottery, tanning and copperware shape the cultural and economic identity of the Medina.

• **Chapter 2: Gastronomy and Culinary Arts - The Flavours of Fez** Immerse yourself in the culinary delights of Fez, renowned for its authenticity and diversity, and explore how Moroccan cuisine enriches the visitor experience.

• **Chapter 3: Accompanying Guides - Revealing Treasures** Explore the crucial role of local guides in providing a deep and immersive understanding of the Medina, their ability to open doors to hidden treasures and enrich the visitor experience.

Exploring the Medina of Fez through its dynasties and historical events is not just a simple tourist visit; it's a real immersion in the glorious past of a city that has both witnessed and played a part in the great moments of Moroccan history. Every step along its winding streets and every encounter with its inhabitants reveals a new facet of this thousand-year-old city, offering travellers an unforgettable experience in which time seems suspended to reveal the hidden treasures of the Medina of Fez.

CHAPTER 1

EXPLORING CRAFTSMANSHIP AS A PILLAR OF EXPERIENTIAL CULTURAL TOURISM IN THE MEDINA OF FEZ

Introduction :

The city of Fez is a popular tourist destination for those looking to explore Moroccan culture. The narrow streets of its medina are filled with artisans working by hand to create unique and traditional products, in spaces dedicated to manual and technical skills. **Experiential tourism** offers a special opportunity for travellers wishing to discover the wealth of local crafts and participate in creative activities in direct contact with the artisans. Traditional crafts abound in the medina of Fez, ranging from pottery and weaving to leatherwork, tannery, copperware and carpentry.These craftsmen use techniques handed down from generation to generation to create unique, authentic products that are appreciated the world over for their quality and beauty. The experience offered to visitors goes beyond simple admiration. The workshops open up to them, revealing the use of ancestral tools such as looms, pottery wheels and leather hammers. Buying artisanal products takes on a new dimension, that of a direct link with those who made them. With this in mind, we'll begin our approach with a historical and cultural contextualisation. We will then explore how visitors are attracted by the opportunity to take part in interactive craft activities, discover traditional trades and immerse themselves in the cultural authenticity of the Medina.Far from being mere spectators, visitors are invited on a sensory journey to the heart of the workshops, where age-old techniques are revealed before their very eyes. They watch as craftsmen dexterously handle tools handed down from generation to generation, the potter's wheel shaping the clay, the loom creating intricate patterns, the hammer striking the leather with precision.Buying a handmade product takes on a new dimension, marked by a deep connection with its creator. Each piece becomes a narrative, a story whispered by the craftsman's expert hands, a vibrant testimony to the soul of Fez.To explore this theme, a rigorous methodology was put in place. Direct observations in the field provided invaluable data on craft trades and tourist activities. Surveys of craftspeople, bazaars and tourist guides provided valuable insights into their perspectives and experiences. Finally, an analysis of tourist testimonials and reviews collected on online platforms such as TripAdvisor helped us to pinpoint visitors' expectations and opinions.

1. The cultural diversity of Fez's population and the variety of the medina's craft structure: the pillars of its craft creativity and productivity

The handicraft sector highlights Morocco's cultural identity, as it has enabled not only the local population to contribute, but also expatriates from Andalusia, Kairouan and other Moroccan regions, who have contributed their own techniques, know-how and experience in various fields. As a result, Fès has built up a large stock of productive assets over the years. As a result, this capital has become a reflection of the originality of Moroccan civilisation, as well as one of its artistic manifestations, attracting tourists from the four corners of the world. In fact, this craft sector has been added to the other monumental and artistic attractions. The aim is to turn the city of Fez into a tourist masterpiece par excellence, an open-air museum. The handicraft sector highlights Morocco's cultural identity, as it has enabled not only the local population to contribute, but also expatriates from Andalusia, Kairouan and other Moroccan regions, who have contributed their own techniques, know-how and experience in various fields. As a result, Fès has built up a large stock of productive assets over the years. As a result, this capital has become a reflection of the originality of Moroccan civilisation, as well as one of its artistic manifestations, attracting tourists from the four corners of the world. In fact, this craft sector has been added to the other monumental and historical attractions to make the city of Fès a tourist masterpiece par excellence, an open-air museum.

The crafts infrastructure is one of the most significant features of the old medina. It contributed effectively to the layout of its alleyways and neighbourhoods, while at the same time making crafts the major activity within it. Although some trades have disappeared, the names of the places where they were practised remain. Some of these districts include the Hadadine district for blacksmiths in Al-Tala'a Al-Kbira, Ain Alou and Al-Nakhline for the manufacture of iron windows, Al-Samarine for cattle plates, Al-Balagine for locks, and so on.

As for the textile and spinning industry and the professions associated with it, there are names such as Al-Harrareen (silk industry), Al-Majadliyine (for caftan belts), Sebaghine for dyeing yarns and dresses, as well as the Selhams and Hayek souks... As for the copperware industry, there is the Seffarine district devoted to the production of metal and copper utensils. As for the leather industry, there are three main tanneries for production:

- **Chouara Tannery:** located on the eastern side of the old town on the north bank of the Oued Boukharab. Thousands of sheep, cow and goat skins are tanned daily to make saddles, babouches and wallets. On the occasion of the national leather fair in 2019, the regional director of handicrafts said of this tannery, "being considered a world tourist destination par excellence, and visited annually by over 90% of tourists coming to the city of Fez. "

- **Sidi Moussa Tannery**: This is located in the Guerniz district near the Mausoleum of Moulay Idris. It covers an area of 1,887 square metres. It contains 85 cupboards and produces tanned cow, sheep and goat leather. This same tannery is also visited by a number of guides and tourists due to its location in the heart of the ancient city. What's more, it and the other tanneries benefited from rehabilitation as part of the programme to restore 27 historic monuments between 2013 and 2017.

Source: fieldwork

- **Ain Azliten Tannery**: located on the northern outskirts, near Al-Talaa Al-Kabira,it is the smallest of the tanneries, with a surface area of no more than 970 metres and 63 cupboards. It specialises mainly in the production of Ziwani leather, which is used in the Ziwani slipper industry[1] .

Just as the names of the districts are related to the types of crafts practised there, they are also linked to commercial activity. Indeed, the names of certain commercial products are given to the whole of the commercial corridor, such as the herbalists (Attarine) where perfumes, medicinal herbs and spices were traded, and Chemaine (sale of candles near the sanctuary of Moulay Idriss).

The Galerie Al-Kifah shopping centre in the heart of the Medina also includes a number of specialist markets arranged in a shopping complex, such as the Jellaba market, the Gold market, the Al-Sabat market (for women's and men's leather shoes), the Al-Harrarin market (silk clothes), the El haïk market (women's fine woollen drapes); the Souk Esselham for burnous (long cape worn by men over the djellaba) and many others.

Figure 2: Plan of the Kissariat AlKifah gallery in the Medina of Fez

Source: Fieldwork

The previous image has been placed on the wall of the different sides of the Al Kifah shopping mall, showing the names of the markets and their distribution in the different zones, serving as a guide for those wishing to wander around. This is a place that stands out for its vast space, multiple wings and elegant organisation. What's more, it has recently undergone restoration and renovation work with a view to restructuring it. It should be noted that the diversity of traditional garments, such as caftans, dresses and Takchita for women, and djellabas, turbans and babouches for men, is one of the most striking symbols of Morocco's cultural identity.In all the medinas, the main thoroughfares are being completely transformed to meet tourist demand, with bazaars and souvenir shops flourishing. The specialisation of streets by activity, which was the original cohesive structure of the Medina, is disappearing, partly because of tourism. The proliferation of shops, craft workshops, fondouks and old mansions transformed into bazaars is most prevalent along the main thoroughfares and at the start of the arterial roads leading off these thoroughfares[2] . In this way, the location of outlets selling handicrafts is transforming the urban space of the Medina thanks to tourism, as they spread out along tourist routes to exploit the flow of visitors as potential buyers. The **bazaar** is a link between cultural tourism and handicrafts, creating a tourist scene in the heart of the old town. Visitors will discover an abundance and variety of products on display, as well as a means of acquiring a personal touch. collection of art objects made by Fassi craftsmen, and an opportunity to perpetuate the memories etched in the city. In this way, the bazaars play a positive role as intermediary commercial entities to promote the diversity of craft products to tourists (by particularly highlighting old products of heritage value, ceramic articles, leather, copper, wood, fabrics... for which the city is famous).

2. Crafts and experiential tourism in the Medina of Fez: a duality between shopping and cultural immersion

While meditating on the Medina of Fez, tourists will notice many artisans hard at work creating handicrafts in their workshops and shops. It is also possible to buy these products directly from these craftsmen or from the shops and bazaars dedicated to handicrafts that can be found all along the tourist routes of the ancient Medina, allowing tourists to bring back a unique and authentic souvenir that reflects their stay in Morocco. This is the motivation behind the vast majority of tourists, who buy objects to remember their trip. Other tourists are motivated by cultural and educational acquisition, as they wish to increase their knowledge and culture. There are also those who buy gifts to give to their loved ones, which is a widespread motivation among tourists, although its intensity varies according to gender, age and nationality.

We note, for example, that the American market has a well-established shopping and consumer culture, as well as significant purchasing power (a fact attested to by most of the guides and players surveyed). For their part, the guides play a central role in introducing tourists to the various craft products that can be purchased when exploring the city, particularly traditional souvenirs. Tourists buy these products according to the way they are presented to them and the need they create in the city. to encourage them to make a purchase. In this respect, the sales assistants employed in the Bazaards can support the guides with heritage explanations relating to the craftsmen's production processes, while trying to persuade tourists to buy the various products on display once they understand how they are made.From our survey of tourists' opinions we found that, in their eyes, the traditional tanneries are the most valuable thing they have seen, and some of them will even go so far as to explain the craftsmen's working methods, and clarify the steps followed after, of course, having properly understood them. From the rooftops, visitors can enjoy a panoramic view of the vats filled with the natural dyes used to dye the leather. It's a unique sensory experience, but it can also be a strong olfactory one because of the scents of the products used.

Figure 3: Artisanal tanneries in the medina of Fez	Figure 4: A tourist comments on the tanneries in the Fès medina
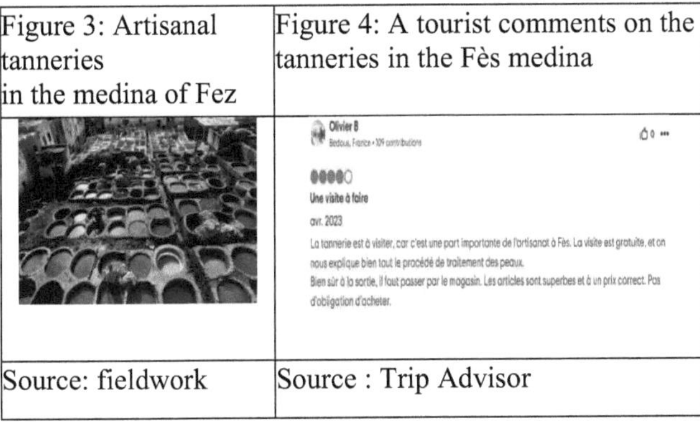	Olivier B Bedous, France • 109 contributions Une visite à faire avr. 2023 La tannerie est à visiter, car c'est une part importante de l'artisanat à Fès. La visite est gratuite, et on nous explique bien tout le procédé de traitement des peaux. Bien sûr à la sortie, il faut passer par le magasin. Les articles sont superbes et à un prix correct. Pas d'obligation d'acheter.
Source: fieldwork	Source : Trip Advisor

A variety of leather products are sold on the various floors, including bags, belts, shoes, babouches, leaf holders, poufs and other typically Moroccan products. The sales assistant explains how Thousands of sheep, cow and goat skins are tanned daily before they become a marketable product, offering tourists the choice of whether or not to buy the various leather products on offer, while at the same time directing tourists to the place where the finished products are displayed[3] .

Source: fieldwork

These include the newly renovated Lalla Yeddouna craft complex in the heart of the old Medina as part of the implementation of the "Crafts-Fès Medina" project, funded by the Millennium Challenge Agency (MCC). The aim of the project was to develop and enhance craft activities, change the urban landscape and create meeting points for craftspeople and tourists as a place to relax and shop, bearing in mind that the square is at the end of the tourist circuit (whether starting from Bab Boujloud or the Andalous quarter) by working to convert it into a space for marketing craft products, including entertainment activities. The site contains various craft activities, such as leatherwork, which involves the manufacture of leather items such as bags and belts, and traditional weaving (Draz), which involves the manufacture of textiles woven using weft. There is also a Foundouk dedicated specifically to the making and display of caftans, which is the art of creating richly decorated traditional women's garments known as 'Dar al Caftan'. Another example is Foundouk Lahssour, dedicated to the leather industry, where you can find craftsmen exhibiting their products and production workshops in their shops.

Figure 6: Plan of the premises at Place Lalla Yedouna Fès

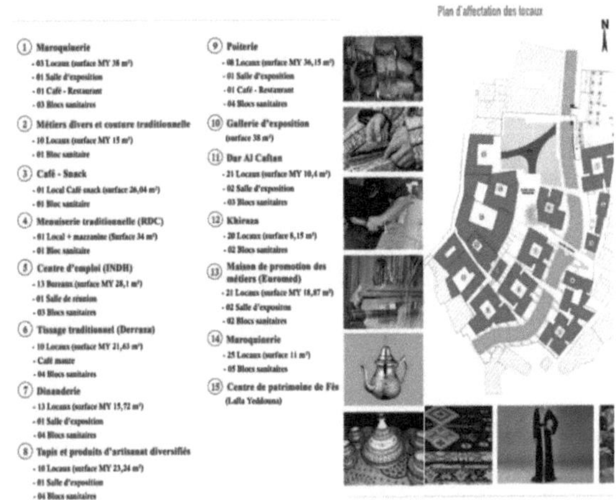

Source : Ader Fès

Figure 7: Foundouk Lahssour in Lalla Yeddouna specialising in leatherwork

Source: Fieldwork

In the Foundouk Lebbata, on the Talaa road, you can still spot a special activity linked to the processing of hides and wool from livestock before going on to the tannery. At the entrance to the Foundouk, there are products on display, and visitors are invited to step inside to observe at close quarters the various stages of production mastered by expert craftsmen. At the entrance you can see how the craftsmen card wool from sheep, goats or cows, using an essential tool called a "wool card". This special brush has fine metal teeth mounted on a flat surface. To smooth the wool, the raw fibre is first prepared by cleaning it to remove impurities.

Figure 8: Wool framing and skin smoothing at Foundouk Lebata Talaa Medina in Fez

Source: Site visit

Then, using the carding machine, the wool is passed back and forth through the teeth, separating tangled fibres, eliminating knots and obtaining more uniform fibres. This step is repeated several times until the wool is smooth, soft and the fibres are parallel to each other. Then you enter the Foundouk area, where another stage takes place to smooth the skin using a tool called a 'Sadrilla' (see figure 8) ... After seeing this process, tourists are encouraged to buy the finished product, which is a carpet made from sheepskin, goatskin or cowhide. These rugs can be used as decorative elements, whether on the floor, on furniture or hung on the wall, adding a warm, aesthetic touch to the interior environment. Pottery, for its part, is showcased in workshops that make up the experience fascinating where the tourists can discover a expertise in the creation of unique, hand-crafted pieces. Visitors have the opportunity to observe the entire manufacturing process in the **potters' quarter in Ain Nokbi** (specialising in the manufacture of traditional pottery and zelliges), from the preparation of the clay to the finishing of the pieces with multiple operations before passing into the hands of decorators who create geometric and floral motifs... During this process they can interact with the craftsmen by asking questions about their craft, technique and history. Groups of tourists are transported by bus every day to visit this site as part of organised tours. After visiting and discovering the authenticity of the products, the craft method and the different stages of production, as well as observing the number of people working in the workshops, tourists are directed to the trading area where they are greeted by local vendors who speak several languages, making it easier to buy these products.

Figure 9: Local mediator explaining the different stages of clay production in the Ain Nokbi potters' district.

Source: Fieldwork

It is still possible to take part in a pottery workshop, as shown by the following comment on TripAdvisor, illustrating the experience of a Norwegian tourist who enjoyed this activity and shared his opinion.

Liah Chen
Oslo, Norvège • 13 contributions

☆☆☆☆☆

Expérience amusante!

C'est une expérience formidable, je le recommande vivement si vous avez plus de temps à passer à Fès. Le guide nous a emmenés à l'usine où nous rencontrons les maîtres et artisans locaux. Il a expliqué la procédure de fabrication des pots en argile, de les colorer et de fabriquer des palettes en céramique. Les maîtres nous ont fait la démonstration pendant qu'il expliquait. Ensuite, nous avons eu la chance de l'essayer par nous-mêmes, avec les conseils des maîtres. On doit ramener à la maison ce qu'on a fait.

Avis sur : Atelier de poterie et de céramique

translated by Google

Écrit le 31 mars 2022

Cet avis est l'opinion subjective d'un membre de Tripadvisor et non l'avis de Tripadvisor LLC. Les avis sont soumis à des vérifications de la part de Tripadvisor.

Voyage65986141719

Thanks a lot for choosing us, we're glad you liked the experience :) we hope to see around next time :)

Figure 10: Coppersmiths at work in the Place Seffarine in Fez

Source: Fieldwork

Pottery products are varied and include tagines, dishes, vases, bowls, plates... Artisans often use bright colours and intricate geometric patterns to decorate their creations, making them attractive and unique objects.

Brassware is another captivating activity in the medina, particularly in the Place **Seffarine.** Passers-by and tourists alike are attracted by the rhythmic sound of hammers, like an improvised concert by traditional craftsmen working in the open air. This lively spectacle, worthy of a most splendid museum exhibition, contributes to Fès' reputation as an open-air museum.There are even more advanced initiatives in terms of experiential craft activities in the Medina. For example, the **Craft Draft** workshops (www.craftdraft.org) initiated by Hamza Fasiki[4] , a young craftsman who comes from a family with a long line of brass players and knew the trade very well. After gaining a master's degree in English cultural studies, he chose to innovate in his parents' business, while at the same time giving himself the mission of passing on and teaching craft skills to locals and tourists, offering them traditional arts and crafts workshops. It's about creating short-term experiences so that people feel like a real craftsman and get an idea of the details of craft techniques involving them making things by hand instead of buying an item that's already made, but with a similar amount it invites them to sit down and make their own product.

Figure 11: Craft Draft introductory workshop for tourists in the Fès medina

Source : Trip Advisor

For example, they can decorate a tray with brass engravings of the five-pointed Moroccan star and embellish it with a number of intricate decorations. In Craft Draft, participants have the opportunity to immerse themselves in a variety of arts and crafts activities. In particular, they can experience bookbinding, where leather is meticulously worked to create unique, personalised covers. The exploration also includes arabesque geometric design[5] , such as Tastir and

Tawriq, techniques traditionally used in art forms such as zellig (mosaic) and pottery. Using a compass, divider, ruler and many other tools, tourists learn to draw, construct and create a variety of original Fez geometric ornaments on paper, plaster, brass, wood and zellige tiles. At the same time, participants also have the opportunity to acquire Arabic calligraphy skills, working with traditional tools such as the Qalam and ink, discovering the art of forming Arabic letters and writing their own names. These experiences have resonated positively with tourists, who rave about them on websites such as Trip Advisor[6] .

The "Draz" weaving workshops offer visitors a unique opportunity to delve into the heart of the fabric creation process. These workshops allow visitors to grasp the precise coordination and skilful technique involved in the harmonious integration of the weft with the warp threads (The "weft" plays a crucial role in the creation of fabrics. Made up of threads woven perpendicular to the warp threads, it creates a harmonious mesh that defines the texture and pattern of the final fabric). After this introduction to the weaving process, they are invited to buy finished products such as shawls in different colours, Hanbals or Jellabas.

Figure 12: Craftsman in a Draz workshop in Place lalla Yeddouna weaving with the weft

Source: fieldwork

CHAPTER CONCLUSION

The Medina of Fez offers much more than a simple walk through the past. It's a living immersion in art and culture that continues through t h e skilled hands of craftsmen. That's why this article has shed light on some of the links between tourism and craftsmanship that enable tourists to experience the Medina of Fez. Nowadays, a whole host of players in the tourism sector are trying to create experiences for tourists, such as strolling through the medina and its souks, staying in a pleasant place like a Riad, tasting a special dish... Craftsmen can now do more than just show tourists what they do, they can sit them down and help them do things by hand. This allows them to experience the work of the craftsmen at close quarters, and to add value to the finished products by acquiring them and thus contributing to the cycle of the tourist economy. By encouraging the development of local crafts through innovative ideas, the city of Fès can benefit in terms of attracting tourists and creating jobs and economic opportunities for local residents. Visitors can not only enjoy high-quality craft products, but also participate in preserving a tradition that is deeply rooted in the city's history and culture.

CHAPTER 2

LINKS BETWEEN TOURISM, GASTRONOMY AND INTERCULTURALITY: COOKING CLASSES AND STREET FOOD IN THE MEDINA OF FEZ

Introduction

The Medina of Fès, a thousand-year-old jewel listed as a UNESCO World Heritage Site, unveils its culinary treasures and invites you on an unforgettable gustatory journey. Its refined gastronomy, the fruit of a rich heritage and multiple influences, offers visitors a unique sensory experience. More than just a tasting session, the cookery classes immerse participants in the heart of local culinary traditions. An authentic immersion that fosters intercultural dialogue and considerably enriches the tourist experience. The boom in culinary tourism is a response to travellers' quest for authentic, immersive experiences. Gastronomy is becoming an essential vehicle for cultural discovery, stimulating the senses and creating lasting memories. Cooking classes are a perfect fit with this trend, allowing visitors to learn the secrets of local cuisine in an interactive way.Far from being simple culinary demonstrations, the cooking workshops, often called "cooking classes", offer an immersive immersion into the world of Fassi gastronomy. They allow participants to interact with local people, discover ancestral culinary traditions and savour authentic dishes prepared with fresh local produce. By focusing on the preparation of emblematic dishes, these workshops enable participants to take ownership of local culinary knowledge and know-how, going beyond simple tasting. This shared immersion fosters a deep understanding of Fassi traditions and lifestyles, creating enriching intercultural links.

1. Gastronomy: A Major Asset for Tourist Attractiveness, Hospitality and Interculturality

The table is the gateway to a destination and a key attraction for tourists. Destinations that highlight their culinary heritage and offer authentic gastronomic experiences have a competitive advantage in the tourism sector. "Moroccan gastronomy has proved that it plays an important role in the choice of destination for a certain category of tourists. Those in charge of the tourism sector need to take this into account and make the most of the opportunities offered by this field, especially as Moroccan cuisine is one of the best in the world"[7] .

Travellers eager for new culinary experiences often turn to Moroccan gastronomy to discover unique and unexpected flavours. According to an interview with an experienced guide in Fez, he confirmed that "tourists are often surprised and impressed by the delicious cuisine of Fez, which reinforces their respect for our culture. They are fascinated by the diversity and quality of the cuisine. The gastronomy of Fez exceeds their expectations and offers them a memorable culinary experience. They recognise that we are a people who can magnificently prepare tasty and balanced popular dishes, without compromising gustatory refinement"[8] . BALAKRISHNAN (2021) goes further, considering that gastrodiplomacy can be considered an edible form of a country's foreign policy.The tourist appeal of the culinary arts is also evident in a number of documentaries specialising in this area, as well as with influencers and tourists who shed light on the richness of this heritage via their videos and vlogs... The richness of Moroccan gastronomy is also reflected in the way hosts welcome their guests. Meals in Morocco are not just occasions for eating, but moments of sharing and intercultural connection. The Moroccan table, always abundant and varied, symbolises the openness and generosity of those who prepare it. When entertaining in Morocco, the importance attached to the meal goes beyond simply tasting delicious dishes.It's a way of celebrating life, sharing joys and traditions, and deepening the bonds between people. The variety of dishes served at a reception reflects the attention paid to the guests, underlining the importance of their presence, "the act of eating is not just a biological necessity, it is also a social and cultural act, encompassing an infinite number of practices that are fraught with meaning (...) Thus Fassi cuisine is not just a matter of flavour, it is staged to feed the eyes, it is given to contemplate"[9] . In short, "the presentation, the choice of particular foods or the selection of stylish cutlery are all elements that define the art culinary"[10] .

2. The Gastronomy of Fez : A Culinary Heritage of Excellence

Cuisine, the very soul of a people, acts like a mirror reflecting ancestral traditions. In Morocco, cuisine is the result of a blend of knowledge and know-how, shaped by the different civilisations and cultures that have left their mark on the country over the centuries. This diversity has transformed Moroccan cuisine into a cultural and artistic mosaic of great value, from the Berbers to the Arabs, via the Andalusians and the Ottomans. Each group has contributed its own culinary traditions, techniques and specific ingredients, creating a unique culinary alchemy.The city of Fez is considered to be the centre of this cultural and culinary heritage, as the capital of Moroccan gastronomy, with a national and international reputation. It is the home of the famous Pastilla, as well as other types of delicious and distinguished dishes, such as Marouzieh, M'hammar, various tajines, khali' al fassi, sweets and couscous. Fès cuisine is a veritable explosion of flavours, spices and aromas. Emblematic dishes such as couscous, tajine, pastilla and harira are prepared with fresh ingredients, carefully selected to offer exquisite combinations of flavours. The use of spices such as cumin, saffron, ginger and cinnamon gives Fès cuisine its distinctive taste signature. Fassi cuisine is characterised by the use of traditional spice blends such as ras el Hanout, and combinations of sweet and savoury flavours. An emblematic example is the pastilla, which brings together a diverse palette of tastes: sweet, salty, sour, meat, nuts and spices, boiled egg... all sprinkled with icing sugar and cinnamon with sheets of filo pastry layered on top enveloping this delicious and unique dish of flavours, which is baked to a crisp texture.

Figure 13: Pastilla, a typical Fassi dish

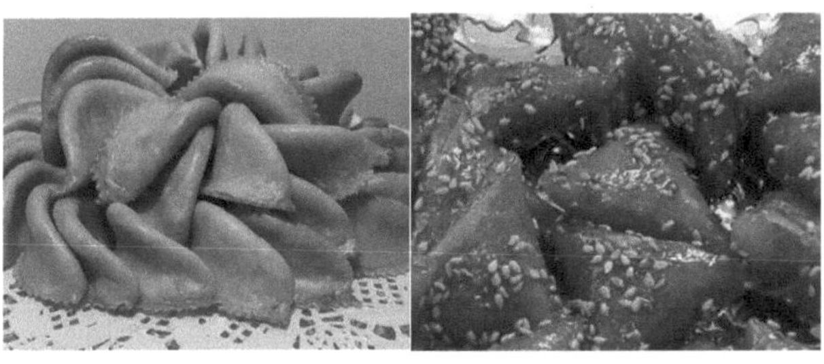

Source : cuisinonsencouleur.com

Moroccan patisserie is a veritable treasure trove of flavour, with its delicious pastries made with almonds, walnuts and honey (Briouates, Chebakias, Cornes de gazelles...). The master patissiers of Fès perpetuate ancestral recipes, and produce creations that are as beautiful as they are delicious. Moroccan pastries are often associated with celebrations and moments of sharing, forming an integral part of the intangible cultural heritage of Fès.

Figure 14: Gazelle horns and almond briouates with sesame seeds

Source : www.saveursdetajine.blogspot.com/2013/06/briouats-aux-amandes.html

The chefs of Fès play an essential role as ambassadors of Moroccan culture through their culinary art. They perpetuate the traditional recipes, ancestral gestures and skills inherited from their grandparents. Passing on this precious knowledge is crucial to preserving the authenticity of Fès gastronomy and sharing it with the world. "In fact, this heritage is passed on through the tales of the old cooks who handle the secrets of this culinary art with ease. These ladies are an indispensable gastronomic encyclopaedia"[11] .

3. Cooking classes as a means of cultural immersion for tourists

Culinary workshops and cooking classes are effective ways of passing on culinary know-how to travellers. These interactive experiences allow participants to learn how to prepare typical dishes under the supervision of experienced local chefs. This is a unique opportunity to experience an immersive experience in Moroccan cuisine, discovering the secrets of Moroccan recipes and leaving with new culinary skills. In a fascinating interview[12] with a seasoned chef in Fez, he shared the immersive methodology adopted in the cooking workshops. From the outset, a warm and friendly atmosphere is established, focusing on the introduction of the participants and a contextualisation of Moroccan cuisine with a description of the cooking equipment, specific utensils and local ingredients. The chef then demonstrates each stage of the preparation process, offering advice on cutting, cooking and presentation techniques. Participants are encouraged to participate actively, receiving individual assistance to ensure understanding and mastery of culinary skills. The emphasis is on sensory exploration, inviting participants to develop their culinary senses by tasting and smelling the ingredients Finally, the convivial moment of tasting the dishes prepared together creates a memorable atmosphere of exchange and sharing. These workshops offer the opportunity to interact with local chefs and allow for direct intercultural exchanges, where participants can ask questions about cooking, daily life and other aspects of

Moroccan culture.On the web, you can find many establishments, guest houses, restaurants and travel agents offering these cookery courses to tourists (at an average price of 55 euros per person), including a tasting of the fruits of their labour.

Figure 15: Comment from a tourist about visiting the souks before the exhibition his cooking workshop

Source : Trip Advisor

Before starting the cooking workshop, many establishments offer visits to the markets and souks of the medina, so that tourists can discover the local ingredients used in Moroccan cuisine and acquire those that will be used in the cooking class. The vibrant and lively atmosphere of the markets creates a captivating sensory experience, offering tourists an immersive insight into local gastronomy.There is also what is known as a Street Food or gastronomic tour to discover the traditional dishes scattered around the alleyways of the Fès medina. Tours include stops at various street food stalls, traditional bakeries, patisseries and local markets. The tastings are adapted to the seasons and to the dietary preferences of the visitors. participants, guaranteeing a personalised experience. From grilled kebabs and sweet and savoury fritters to traditional soups, Bissara (made from beans) and Moroccan pastries (briwa, chebbakia, etc.). Tourists have the opportunity to sample authentic dishes prepared before their very eyes.

4. Gathering tourists' opinions on Cookings Classes

To shed light on this area, an analytical study was carried out, based on an analysis of tourist reviews published on Tripadvisor. This analysis provided an insight into visitors' perceptions of Moroccan gastronomy, focusing on cooking workshops, gastronomic tours and visits to local restaurants. According to feedback from the TripAdvisor.com website, Fassi's gastronomy is attracting a great deal of interest from tourists in search of the region's culinary specialities.

Their gastronomic experience with Cooking Class testifies to their satisfaction with a cuisine that they will discover, carefully and gradually, with the local people, who have generously opened their doors to them.One American traveller said: "During my cooking workshop in Fez, I learned how to prepare a delicious tajine. It was an authentic experience that allowed me to immerse myself in Moroccan culinary culture and understand the subtleties of the local spices and ingredients"[13] .

The tajine, an emblematic Moroccan dish, chicken with olives and preserved lemons, meat with prunes and almonds, or the vegetable tajine, stands out thanks to the distinguished appreciations and approvals of tourists. Also, Fès couscous, very popular in Morocco, served with vegetables and meat or chicken, leaves a lasting impression on visitors.Pastilla, a sweet and savoury speciality made from sheets of brick pastry, chicken or pigeon, almonds and icing sugar, appeals to tourists with the most refined tastes. Last but not least, there are the regional pastries, which make people's mouths water when they discover them for the first time. They are made with almonds, honey and cinnamon. The best known of these are gazelle horns, Briouates with almonds, makrouts and Chebbakias.

Figure 16: Example of dishes offered in the cooking classes of a restaurant in the medina of Fez

https://www.cafeclock.com/cooking-classes-1

- Zalouk - roasted aubergine with spices
- Khizou M'shrmel - marinaded carrot salad
- Shlada d l'barba - beetroot salad
- Taktouka - spiced green pepper salad

Soups, a choice of:

- Harira – traditional hearty Moroccan soup
- Bissara – split pea or broad bean soup with spices, garlic and olive oil

MAINS (Couscous, Tagine or B'stella) ,a choice of:

- Seksou b Sbeع khdari - Seven vegetable couscous
- Seksou T'faya - Couscous Boohaloo: with almonds, caramelized onions and raisins
- Tagine b t'mer w l'berkok - Tagine with prune and date
- Tagine b l'khodra: seasonal tagine with vegetables such as quince & okra or wild with a choice of Chicken, Lamb or Beef
- Tagine djaj m'kalli - chicken with preserved lemon and olive
- Tagine b l'hout - fish tagine (dependent on availability in souk)
- B'stella b'djaj – chicken bastilla
- B'stella b l'hout – fish bastilla

DESSERTS a choice of:

- Letshine b l'karfa - orange & walnut salad
- Ghriba d l'3sel - honey fassi maccaroons

Source: cafeclock.com/cooking-classes-1

Now, having carefully analysed the comments made by international tourists about their culinary experience on the Fassi terrain, it is clear that the explorers were unanimously satisfied with "(Super experience that we would recommend.), (Thank you for a great time), (Extraordinary experience) (Fantastic authentic Moroccan culinary experience)...". While some were astonished by the potential variety of culinary methods: "You're going to enjoy food in a different way, by discovering new techniques", others found more pleasure in demystifying the rites of the souk and its secrets: "Abou, who introduced us to the souk, its codes and its specialities, delighted our taste buds...".

Source : Trip Advisor

Participants in cooking classes are motivated by both social needs such as the desire for social interaction and utilitarian elements such as the acquisition and application of practical skills. On the experience side, participants are looking for a mix of entertainment and escapism, seeking both pleasure and an escape from the daily routine. The educational aspect of the cooking classes seems to have been particularly striking for the participants, suggesting that these activities are not simply entertainment, but also opportunities for meaningful learning. The entertainment element, meanwhile, helped to create an overall enriching experience. Many of these results have also been demonstrated in other tourist areas where these workshops are offered (SUNTIKUL et Al, 2020; KOKKRANIKAL and CARABELLI, 2021; ARTUKLU, 2022). In short, cooking classes offer memorable experiences, immersing participants in the richness of Moroccan cuisine. These moments combine the pleasure of cooking with an intercultural immersion, creating gustatory memories to keep tourists happy.

CHAPTER CONCLUSION

The tourism of intangible heritage, in particular gastronomy, is a unique opportunity to showcase a thousand-year-old culinary tradition and promote interculturality. Gastronomic tourism highlights the talent of local restaurateurs and chefs and their contribution to preserving the city's culinary traditions. Fès gastronomy is much more than a simple taste experience, it's a journey through time, an immersion in the history and culture of an ancestral city. Authentic culinary experiences can help to retain tourists and recommend the destination to other travellers. Visitors who have enjoyed a unique culinary experience are more likely to return to the destination in the future, or to recommend it to friends and family. Gastronomy can therefore play an important role in promoting and retaining tourists.In short, tourists' amazement at the culinary delights reinforces their respect for our culture and civilisation. The gastronomy of Fez is a showcase of culinary heritage and creativity, and the fact that visitors do not expect such a culinary experience reinforces their appreciation of our culture. By promoting the gastronomy of Fez, the city can attract quality culinary tourism, contributing to the growth of the tourism industry and the reputation of the destination.

CHAPTER 3

"ACCOMPANYING GUIDES: CULTURAL AMBASSADORS FOR HERITAGE INTERPRETATION AND CULTURAL IMMERSION IN THE MEDINA OF FEZ"

Introduction

The Medina of Fez attracts thousands of tourists every year, drawn by its rich history and heritage, which sums up the different stages of Moroccan history over the last 12 centuries. As one of the oldest Islamic and Arab capitals in North Africa, it is a great opportunity to step back in time and discover the many facets of Moroccan history and culture.However, exploring the Medina of Fez can be difficult for visitors, due to the complexity of its narrow streets (3,000 alleyways) and labyrinth of historic buildings (11,000 buildings of different types: 740 palaces and beautiful residences, 176 mosques, 83 mausoleums and zaouias, 11 medersas...). Visitors can easily get lost or miss important sites during their visit, which can limit their cultural experience.

Semi-structured interviews were carried out with a sample of tour guides who had extensive experience in the sector and in-depth knowledge of the Medina of Fez. This enabled us to gather in-depth information on the issue in question, revealing the experiences, perspectives and challenges encountered by tour guides in their role as intercultural facilitators.

Periodic visits were also made to the Medina of Fez to accompany the guides on their tours of the Medina and to see at close quarters their contribution to the presentation of the heritage sites and the promotion of craft products, as well as guiding tourist itineraries through the old streets of the medina.

1. Multiple roles for guides

According to the Fez Tourism Delegation, by 2023 the number of official guides had risen to 560. These professionals are paid according to various criteria, such as the length of time they accompany tourists, the number of tourists they guide, their skills and their experience in the market. Rates generally vary between 250 and 800 dirhams per day, depending on the services provided. Guides may also receive tips from tourists as a reward for quality service. Tourist guides play a crucial role in the visitor experience by revealing the rich heritage of the Fès medina. By engaging in an interactive dialogue with tourists, they provide a better understanding of the historical significance of the site and encourage appreciation of its cultural dimension.As cultural facilitators, local guides in Fez help overcome language barriers and encourage interaction between travellers and locals. Their presence makes it easier for visitors to take part in authentic local activities, enriching their tourist experience.Fez tour guides also act as cultural ambassadors for the city, providing tourists with in-depth information about the cultural heritage, history and local customs. Their accompaniment to tourist sites and historic monuments allows visitors to immerse themselves in the history and social life of the city. The positive impact of a successful guide service is often reflected in the satisfaction of tourists, who express their delight at discovering the hidden secrets of the medina thanks to the expertise of the guides. One tourist underlined the importance of this experience by saying: "Visiting the medina with a guide is, in my opinion, essential for understanding all its history and secrets and discovering places that are almost hidden". Tour guides must have a wide range of skills, including an in-depth knowledge of local culture, history and traditional crafts, as well as the ability to respond to tourists' questions and needs. They generally spend between 5 and 8 hours a day on the tour, and can be even longer on city-to-city tours. The medina of Fez, in particular, is a labyrinth of narrow streets and intricate mazes. The guides know the best routes, the must-see attractions and the lesser-known but equally fascinating places. As such, they play an important role in helping travellers navigate the complexities of Fez.Our study, based on the accounts of 120 international tourists who shared their adventures in the medina on TripAdvisor, reveals a profound dimension of experiential tourism. Among these travellers, 39% expressed a common concern: getting lost in the intricacies of the old town's labyrinthine alleyways. It was in this context that the intervention of an official tourist guide was particularly valued. For these visitors, hiring a certified guide was much more than just a safety measure. As well as navigating smoothly through the historic districts, the guide played a crucial role in enriching their cultural experience. Thanks to his in-depth local knowledge, he was able to not only only to illuminate their exploration by revealing hidden aspects of the medina, but also offer them valuable insights into the history, architecture and life local newspaper. Interaction with an accredited guide enabled these travellers to immerse themselves authentically in Moroccan culture, with complete peace of mind. They were able to discover local traditions, taste traditional cuisine and interact with local people, enriching their stay with memories and profound learning.

2. Orientation of tourist circuits in the Medina of Fez and cultural immersion in its alleyways

The guides' main task is to plan and direct tourist routes in the Medina based on their experience in the field, taking into account various criteria such as the motivations and interests of the tourists being guided, their physical ability to wander through the narrow streets of the Medina and the time available for the visit. From our field surveys with tourist guides, we found that the tours they provide in the Medina, most of which take one day (bearing in mind that the average stay in Fez is two days[14]), are divided into different circuits, with a distinction being made between extra-Muros and intra-Muros tours.

- Extra Muros circuit (map 1): this is done using tourist transport, starting from the royal palace and its ornate gateway next to the Mellah district. Then head to the Borj Sud next to Bab Ftouh for a view of the city.fascinating panoramic view of the Medina and its monuments such as the minarets of the Qarawiyin and Andalusian mosques ...

The tour then continues to the Fakharine Ain Nokbi district to discover the pottery workshops. The tour can also take you to the Borj Nord, built during the Sadian era and home to the Weapons Museum, which offers another panoramic view of the whole of the old Medina.

- Intramural routes (map 2): three main routes used by guides with tourists in the old Medina can be identified, with visitor numbers varying as follows:

A. Bab Boujloud - Oued Zhoun circuit: (30% use by guides surveyed)

The entrance to this tour is easily accessible by car from the famous Bab Boujloud gate (known for its beautiful architecture and the green and blue ceramics on both facades), then down two main thoroughfares Talaa Kbira or Sghira[15] towards the centre of the Medina and where you can discover the main monuments of the Medina such as :

- Medersa Bouanania and Medersa Attarine

- Foundouk, fountain and place Nejjarine , place Seffarine

- Tannerie Sidi Moussa / Tanneries Chouara

- Historic triangle: Moulay Driss Mausoleum / Quaraouiyine Mosque (after

its Chemmaine or Bab Lwerd gate) and Zaouia Sidi Ahmed Tijani.

The historic triangle concentrates craft and commercial activities (Kissarias, souks, foundouks, tanneries, dye works, brass works, etc.) and attracts large numbers of visitors, both national and foreign, as it is located in the heart of the medina and is dotted with the landmark monuments of Fez, as shown in the figure below.

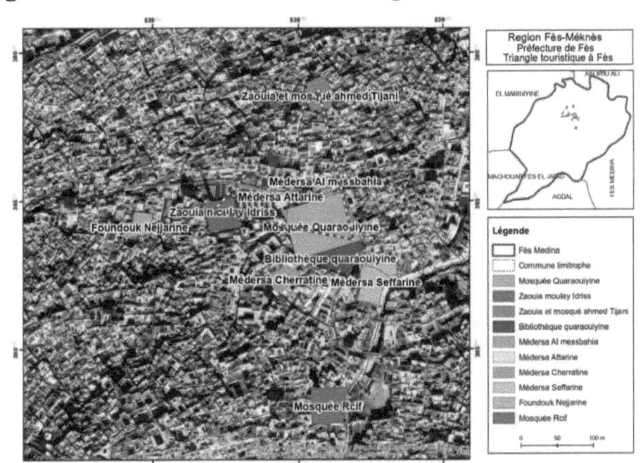

Figure 18: Axes of the historic triangle in the Medina of Fez

Source: Satellite image (Esri)

In most cases, this circuit leads to the Oued Zhoun exit which has a car park.

B. Rcif-Oued Zhoun circuit: (60% use by guides surveyed)

This tour is characterised by its **easily accessible topography**, with access via the Rcif entrance and its Sid el Aouad Gate, allowing visitors to discover the main historical monuments in the centre of the Medina in a very short space of time, which explains why it is so popular with most guides as well as the first tour:

-Place Seffarine and Quaraouiyine Library

- Quaraouiyine Mosque (its Bab el Ward or Bab Chemaine Gate)

- Medersa and Souk Attarine (the Cherratine medersa can also be visited)

- Fountain, Place and Fondouk Nejjarine

- Moulay Driss Mausoleum and Sidi Ahmed Tijani Zaouia

- Tannerie Chouara and exit via Oued Zhoun

C. Andalusian Quarter Tour: (10% rate of use by guides surveyed)

This route is known for its small number of tourists[16] compared to other routes[17] , although a number of guides take it with tourists to show them another aspect of simplicity and popular life, and for Spanish tourists because of its connection with the Andalusians (from whom it takes its name) who settled in this area after emigrating...This trail often starts at Bab Al-Khoukha before visiting the two main landmarks of this Quarter, namely the Al-Andalus Mosque and the the marvellous Medersas Al-Sahrij nearby (transformed into a calligraphy centre after its restoration). With the inauguration of the craft complex project in Place Lalla Yadouna, it has also become a tourist resort before reaching Place Seffarine by taking Rue Mashatine and ending the journey at Rcif.

Map 1: Extra muros tourist circuit of the Medina of Fez

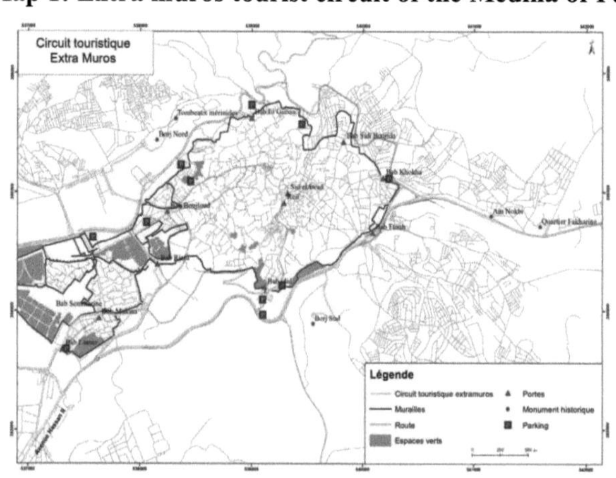

Source: Field survey

Map 2: Main routes taken by tourist guides in the Medina of Fez

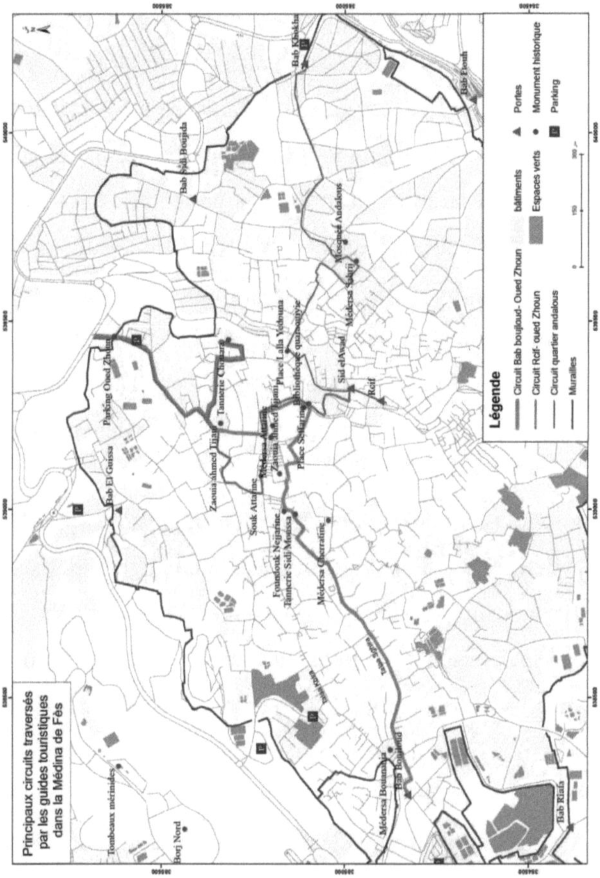

Source: Field survey

3. Tourist Guides and the Weaving of Intercultural Dialogue

The tour guide plays a crucial role in raising awareness of cultural heritage by exploring various itineraries and presenting different historical monuments. An experienced guide, used to dealing with tourists of different nationalities (notably from England, America and Australia), emphasised the rich heritage of the city of Fez and that he cannot therefore explain all these aspects to visitors, which forces him to condense the information, making it captivating to hold attention and arouse the desire for discovery. He also noted that tourists are amazed by the rich heritage of Fez, which bears striking similarities to Andalusian civilisation, as witnessed by the Madrasa Al-Attarine and the Bou Inania, reminiscent of the sumptuous Alhambra palace in Granada.Another English-speaking guide said that most tourists prefer Fez to other cities in the kingdom. He showed us testimonials from international tourists he has accompanied, and how his expertise and professionalism have had a great impact on them. In their comments, they expressed their preference for Fez over other cities such as Rabat, Casablanca and Marrakech. The guide also works to "promote the acceptance of cultural differences as an essential basis of human diversity, and to reinforce the presence of certain cultural symbols of the host country among tourists, whether through certain linguistic expressions, cultural gestures, or by wearing traditional clothing and encouraging tourists to buy it. It strives to correct negative stereotypes about the host society in the minds of tourists and to dispel misunderstandings about certain religious notions in the host country, such as the perception of Islam in the West". This can be used as publicityindirectly positive for other tourists, by giving the guide the opportunity to to present a correct image in response to distorted images in the media, and to counter ignorance of the Arab and Islamic world, its civilisation and culture. At the same time, tourist guides need to be trained to dialogue and interact with international tourists on religious issues, as some tourists arrive with misconceptions and questions that require wise and informed answers (such as questions relating to women, the veil, violence, etc.). For example, to answer the question of the inferior place of women in Islam, he cited the example of the city of Fez, where Al Quaraouiyine University, considered to be the oldest university in the world teaching various religious and secular sciences, was founded by a woman, Fatima Al-Fihriya. He also points out that the place of women in the West was not very important either, since they only recently obtained the right to vote (in 1944 in France, for example).Another Spanish-speaking tour guide stressed that treating tourists humanely is essential to leaving a positive impression. Indeed, many comments on TripAdvisor praise his good guidance and good behaviour, and recommend taking advantage of his services. He stressed the importance of presenting Arab-Islamic culture in an attractive way and showing how it interacts with other cultures. He is prepared to adapt to all types of tourists with dignity, wisdom, a smile and hospitality, and, if necessary, with firmness when it comes to matters of national and religious sovereignty, while working to correct misinformation and preconceived ideas by providing accurate information. To do his job in the best possible way, he spares no effort in continuing education, reading and discussing with experts to keep up to date. Indeed, tourists often come with a thirst for knowledge and prior information obtained from various platforms in online, requiring the guide to have more extensive, precise and detailed information.Another tour guide shared an experience with a British tourist, writer and author. Although the latter had read several books about Morocco and the city of Fez prior to his visit, he recognised that the information provided by the tour guide was unique and distinctive, and that he had not found it in the sources he had consulted previously. Tour guides also facilitate a time-travel experience by allowing tourists to relive past eras through local stories and anecdotes. As players in experiential tourism, they allow visitors to Fez to immerse themselves in the rich cultural and historical fabric of the city, fostering a deeper and more immersive understanding of local culture. This cultural immersion plays an essential role in intercultural dialogue, enabling tourists to understand and appreciate the differences and similarities between their own cultures and those of Fez. In this way, the guide becomes a mediator, forging links between visitors and the local community, and encouraging an enriching exchange of ideas and perspectives.

4. Tourist guides: key players in the interpretation of cultural heritage

Tourist guides play a crucial role in guiding visitors' itineraries through the medina of Fez, presenting the city's various heritage treasures. They take into account a number of criteria, including those highlighted by the majority of guides, such as visitors' motivations and specific interests, as well as their physical condition, age and the length of their visit. Guided tours through the medina of Fez offer a variety of routes to discover different facets of the city's heritage. Among these itineraries, the guides highlight various historic monuments, for example :

- **The Bou Inania Medersa:** Medersas proliferated during the era of the Marinids, at the height of civilisation, who sought to embellish these establishments to provide a favourable environment for students. Founded to train the future elites of the state, they attracted students from all over Morocco. Today, these medersas captivate tourists from all over the world, who come to discover their beauty and the traditional crafts they house. Guidebooks try to include one of these medersas on their tours to highlight the scientific and cultural importance of Fez.

The Bou Inania Medersa is one of the most famous and most visited in Fez. Built by Sultan Abu Inan Merinid, it bears his name and is renowned for its artistic creations in ceramics, plaster and wood, in harmony with the architecture and artistic civilisation of the Merinid era within its walls. The traveller Ibn Battuta said of it: "I have seen nothing like it in Syria, Egypt, Iraq or Khurasan". With its two floors and 40 rooms, the medersa is also the only place where Friday prayers are celebrated in addition to the five daily prayers. In addition, there is a water clock in front of the school that highlights the technological development that reached the city during the era of the Merinids of Morocco.

Figure 19: The courtyard of the Bou Inania medersa and its water clock

Source: Personal photograph

It should be noted that these institutions, in addition to other historic monuments, host intellectual meetings and conferences, notably as part of the Sufi Culture Festival or the Sacred Music Festival, underlining the importance of cultural dialogue between peoples.The Festival of Sacred Music, created in 1994 on the initiative of Fawzi Squali, an influential figure in the fields of anthropology and Sufi thought, in response to the events of the Gulf War, aims to make an artistic contribution to encouraging rapprochement between peoples and promoting the culture of peace (two important conditions for developing international tourism and avoiding crises). Since 2001, the festival has been selected by the United Nations as one of the most important events contributing significantly to the dialogue of civilisations, gaining international renown thanks to various international and national audiovisual and written media.

Figure 20: Intellectual conference on the sidelines of the Sufi Culture Festival at the Bou Inania Medersa

Source: Personal photograph

Figure 21: Artistic evenings at the Music Festival spiritual space at Jinan al-Sabeel

Source: Personal photograph

The festival makes a significant contribution to promoting the city of Fez and enhancing its appeal as a tourist destination. Each year, a specific theme centred on dialogue, peace and closeness between peoples is selected and discussed throughout the days of the festival at round tables and parallel conferences. These discussions are led by a team of researchers and academics from Morocco and abroad, attracting tourists of a high cultural level.The festival offers a unique opportunity for artistic encounters and exchanges between the world's civilisations and cultures. Each year, the festival welcomes a number of renowned artists from all over the world, giving the public the chance to discover the artistic productions of various countries.

CHAPTER CONCLUSION

In the medina of Fez, heritage interpretation is crucial to bringing its rich history and vibrant culture to life. Tourist guides play a central role in this endeavour, revealing the profound significance of the iconic sites, age-old traditions and craft practices for which the city is renowned.Recognising the importance of guides and supporting their ongoing training are essential aspects of strengthening the positive impact they can have on cultural tourism in Fez. By investing in their professional development, we ensure a higher quality of heritage interpretation, which creates memorable experiences for visitors. Successful heritage interpretation not only enlightens visitors about the Medina's glorious past, but also stimulates their appreciation of the city's present. By understanding the stories and traditions passed on by the guides, tourists can gain a better understanding of the cultural and historical richness of Fez, enriching their travel experience and leaving them with lasting memories.

CONCLUSION OF THE BOOK

Heritage interpretation in Fez is crucial to bringing to life the rich history and vibrant culture of the Medina. Guides play a central role in revealing the deeper meaning of emblematic sites, age-old traditions and craft practices.Some guides encourage visitors to take an active part in craft demonstrations. Whether it's pottery making, coppersmithing, or other traditional crafts. This hands-on immersion allows visitors to understand the creative process behind the crafts, enhancing their appreciation of these trades. Successful interpretation of heritage creates memorable experiences, stimulating appreciation of the Medina's past and present.The immersive guided discovery of craft and gastronomic treasures offers a complete and enriching experience, allowing visitors to connect deeply with local culture. By taking part in workshops, visiting artisans' studios, taking part in cookery classes or sampling local products, travellers can not only appreciate the riches of a region but also contribute to the preservation of its living heritage. This approach enhances the tourist appeal of the regions while supporting local craftsmen and producers.The immersive guided tour of the Fès medina goes beyond a simple stroll through the historic streets; it's an interactive journey through the ages. Combining living history, impressive architecture, traditional craftsmanship and culinary delights, this experience allows visitors to feel, touch and taste the richness of Fez's past. It is a unique way of understanding the city, not just as a geographical location, but as a living witness to Moroccan history and culture. Tourist guides play an essential role in passing on knowledge about the culture and heritage of the city of Fez. Accompanying tourists to tourist sites and historic monuments, they enrich their experience by providing detailed information about the cultural heritage, history, customs and social life of the local people. Thanks to their local expertise, they facilitate an in-depth and enlightening exploration of Fez, offering visitors an enriched perspective and lasting memories of their stay in this historic Moroccan city.

BIBLIOGRAPHY

- **Aziz Hmioui, Amina Haoudi 2016:** The role of gastronomy and handicrafts in the tourist appeal of the city of Fez: a study based on the perceptions of foreign tourists. Revue Management & Avenir /3 (N° 85), pages 149 to 169.

- **Berriane Mohamed 1980,** L'espace touristique Marocain. Fascicule de Recherches N°7, Centre interuniversitaire d'études Méditerranéennes- Poitiers, Publication ERA N°706, Imprimerie de l'Université de Tours, France.

- **Bipithalal BALAKRISHNAN 2021**, "Gastrodiplomacy in Tourism: Capturing Hearts and Minds through Stomachs". International Journal of Hospitality & Tourism, Volume 14, Issue 1, Publishing India Group, p30-40.

- **Hassan FAOUZI 2018**, "Virtual communities: the power of information in the world of gastronomy, the case of restaurants in Agadir Morocco". In Tourism, governance, ICT and territorial policy in Africa. Agadir. International University of Agadir, p83-105.

- **Jéssica Ferreira & Bruno Sousa 2019**, Experiential Marketing as Leverage for Growth of Creative Tourism: A Co-creative Process / Part of the book series: Smart Innovation, Systems and Technologies (SIST, volume 171) Advances in Tourism, Technology & Smart Systems Springer.

- **Jithendran KOKKRANIKAL, Elisa CARABELLI 2021**, "Gastronomy tourism experiences: the cooking classes of Cinque Terre". TOURISM RECREATION RESEARCH. https://doi.org/10.1080/02508281.2021.1975213.

- **Mardin ARTUKLU 2022**, "Is it possible to get to know a culture through cooking classes? Tourists experiences of cooking classes in istanbul". International Journal of Gastronomy and Food Science ,Volume 28, https://doi.org/10.1016/j.ijgfs.2022.100527.

- **Muriel Girard 2004,** Mise en scène touristique de l'artisanat traditionnel dans le contexte de patrimonialisation de la médina de Fès : les cas de la tannerie

Chouara et du quartier des potiers, Afemam workshop, CITERES-EMAN laboratory, François Rabelais University, Tours.

- **Salah Chakor 2008,** Traité de gastronomie marocaine, Tangiers, Imprimerie le journal.

- **Samira FORAJ 2023**, " à la découverte de l'art culinaire marocain ", Mon Livret du Nord du Maroc, Ping Pong Magazine, November issue, http://www. drive.google.com/file/d/1T110HUOBiPfkqttSpPkCcmACRazkDm8Q/view

- **Sofia EL MOKRI 2008**, "La cuisine Fassie une réalité sublimée, regard d'une bourgeoisie sur elle-même". In Horizons Maghrébins n°59. Mirail University Press.

- **Wantanee SUNTIKUL, Elizabeth AGYEIWAAH, Wei-jue HUANG, AND STEPHEN PRATT 2020**, "Investigating the tourism experience of Thai cooking classes: an application of lasern's three stage model". Tourism Analysis vol 25, ttps://doi.org/10.3727/108354220X15758301241684.

- Cooking classes in Fez - Tripadvisor www.tripadvisor.fr/Attraction_Products-g293733-t12034-zfg11868-Fes_Fes_Meknes.html

- www.craftdraft.org

TABLE OF CONTENTS

I want morebooks!

Buy your books fast and straightforward online - at one of world's fastest growing online book stores! Environmentally sound due to Print-on-Demand technologies.

Buy your books online at
www.morebooks.shop

Kaufen Sie Ihre Bücher schnell und unkompliziert online – auf einer der am schnellsten wachsenden Buchhandelsplattformen weltweit! Dank Print-On-Demand umwelt- und ressourcenschonend produziert.

Bücher schneller online kaufen
www.morebooks.shop

Printed by Books on Demand GmbH, Norderstedt / Germany